CONFERENCE REPORT

WINNING THE WAR BY WINNING THE PEACE: STRATEGY FOR CONFLICT AND POST-CONFLICT IN THE 21st CENTURY

Fifteenth Annual Strategy Conference
Carlisle Barracks, Pennsylvania
April 13-15, 2004

Lloyd J. Matthews

December 2004

*****

The views expressed in this report are those of the author and do not necessarily reflect the official policy or position of the Department of the Army, the Department of Defense, or the U.S. Government. This report is cleared for public release; distribution is unlimited.

*****

Comments pertaining to this report are invited and should be forwarded to: Director, Strategic Studies Institute, U.S. Army War College, 122 Forbes Ave, Carlisle, PA 17013-5244. Copies of this report may be obtained from the Publications Office by calling (717) 245-4133, FAX (717) 245-3820, or by e-mail at *SSI_Publishing@carlisle.army.mil*

*****

All Strategic Studies Institute (SSI) monographs are available on the SSI Homepage for electronic dissemination. SSI's Homepage address is: *http://www.carlisle.army.mil/ssi/*

*****

The Strategic Studies Institute publishes a monthly e-mail newsletter to update the national security community on the research of our analysts, recent and forthcoming publications, and upcoming conferences sponsored by the Institute. Each newsletter also provides a strategic commentary by one of our research analysts. If you are interested in receiving this newsletter, please let us know by e-mail at *SSI_Newsletter@carlisle.army.mil* or by calling (717) 245-3133.

ISBN 1-58487-182-2

FOREWORD

During each of the last 15 years, the U.S. Army War College has sponsored a broad-based strategy conference that addresses a major security issue of current relevance to the United States, its allies, and, indeed, the entire world. Bringing together some 150-200 scholars, defense specialists, news media representatives, active and retired members of the military community, and uniformed and civilian faculty members of senior U.S. service colleges, the annual conferences are conceived to generate open, unfettered dialogue on the issue under discussion. Particular attention is paid to controversial or unresolved questions, always with the aim of surfacing practical answers based upon multiple perspectives and a dispassionate regard for reason and the attendant facts.

The conference theme for year 2004 was "Winning the War by Winning the Peace: Strategy for Conflict and Post-Conflict in the 2st Century." Informed by the Korean, Vietnam, and Gulf wars fought by the United States and its allies during the last half of the 20st century—wars in which, despite the qualitative superiority of our forces, the outcomes proved to be less than satisfactory—the conference theme for 2004 entailed a deep probe into the question of how can the West, in this new century of omnipresent terrorism, capitalize on its superior military and economic might to achieve a satisfying and enduring *modus vivendi*. The search for answers to this central question was lent added relevance and urgency by the fact that the allied anti-insurgency wars in Afghanistan and Iraq were transpiring even as the conference proceeded and, indeed, even as this report goes to press.

The Strategic Studies Institute is pleased to provide this summary of the 2004 conference presentations along with a distillation of conference findings and conclusions.

DOUGLAS C. LOVELACE, JR.
Director
Strategic Studies Institute

BIOGRAPHICAL SKETCH OF THE AUTHOR

LLOYD J. MATTHEWS is a retired U.S. Army colonel. His military assignments included command at platoon, company, and battalion level; advisory duty in Vietnam; editorship of *Parameters*, the Army War College quarterly; and the associate deanship of the U.S. Military Academy. Following retirement from the Army, he was a project manager in Saudi Arabia and Turkey. Colonel Matthews is the author of over 100 articles, features, reviews, and monographs on professional topics. Most recently, he contributed the entry on American military ideals appearing in *The Oxford Companion to American Military History* (Oxford University Press, 1999), and was coeditor, along with project directors Don M. Snider and Gayle L. Watkins, of *The Future of the Army Profession* (McGraw-Hill, 2002). A second edition is forthcoming in 2005. Colonel Matthews has a B.S. degree from the U.S. Military Academy, an M.A. from Harvard University, and a Ph.D. from the University of Virginia. He is also a graduate of the Armed Forces Staff College and the U.S. Army War College.

CONFERENCE AGENDA

Tuesday, April 13

11 a.m.-3 p.m.	Pre-Conference Elective #1: Gettysburg Staff Ride
4:30 p.m.-7 p.m.	Pre-Conference Elective #2: *The Battle of Algiers* Movie Discussion

Wednesday, April 14

8:30 a.m.-10 a.m.	Keynote Address, LTG Jay Garner, U.S. Army (Ret.), former Director of the Office of Reconstruction and Humanitarian Assistance in Baghdad
10:30 a.m.-12 noon	Panel I: *American Grand Strategy in the Global War on Terrorism*

Chair: COL John Martin, Deputy Director, Strategic Studies Institute, U.S. Army War College

Panelists:

Prof. William C. Wohlforth, Dartmouth College
Prof. Robert Lieber, Georgetown University
Dr. Robin Dorff, Chairman, Dept. of National Security and Strategy, U.S. Army War College

1:45 p.m.-3:15 p.m.	Panel II: *Fiscal and Political Sustainability*

Chair: Dr. Stephen Biddle, Strategic Studies Institute, U.S. Army War College

Panelists:

Dr. Michael O'Hanlon, The Brookings Institution
Dr. Eric Labs, Congressional Budget Office
Dr. Alice Rivlin, The Brookings Institution

3:45 p.m.-5:15 p.m. Panel III: *Insurgency and Terrorism*

Chair: Prof. David Haglund, Queens University

Panelists:

Dr. Steven Metz, Director of Research, Strategic Studies Institute, U.S. Army War College
Prof. Ian Beckett, Horner Professor of Military Theory, U.S. Marine Corps University
Mr. Steven Simon, RAND

6:30 p.m.-8:30 p.m. Banquet Presentation, Mr. William Kristol, Editor, *Weekly Standard*

Thursday, April 15

8:30 a.m.-10 a.m. Panel IV: *Peacekeeping, Nation-Building, and Stabilization*

Chair: COL Michael J. Dooley, Director, U.S. Army Peacekeeping and Stability Operations Institute

Panelists:

Dr. Conrad Crane, Director, U.S. Army Military History Institute
Ms. Bathsheba Crocker, Center for Strategic and International Studies
Mr. Jaque Grinberg, Chief of Staff to Special Representative of the Secretary General for the Democratic Republic of the Congo

10:30 a.m.-12 noon Panel V: *Democratization*

Chair: Dr. Andrew Terrill, Strategic Studies Institute, U.S. Army War College

Panelists:

Prof. Daniel Brumberg, Georgetown University
Dr. Larry Diamond, Stanford University
Prof. Andrew Reynolds, University of North Carolina, Chapel Hill

12:15 p.m.-1:30 p.m. Luncheon Presentation, Dr. Lawrence J. Korb, Former Assistant Secretary of Defense (Manpower, Reserve Affairs, Installations, and Logistics)

1:45 p.m.-3:15 p.m. Panel VI: *Implications for Military Capabilities and Force Structure*

Chair: Prof. Douglas Lovelace, Director, Strategic Studies Institute, U.S. Army War College

Panelists:

Dr. Daniel Goure, Lexington Institute
COL Richard Hooker, Planner, Army Staff
LTC Robert Cassidy, Member, U.S. Army, Europe, Commanding General's Initiatives Group

3:30 p.m.-4:30 p.m. Briefing, BG Kevin T. Ryan, Director, Plans and Policy, Office of the Deputy Chief of Staff, G-3

INTRODUCTION

Colonel Lloyd J. Matthews, U.S. Army Retired

With little doubt, three of the most pressing and frequent problems grappled with in Western defense and geostrategic literature over the past 20-30 years have been how to fight asymmetric wars, how to win the hearts and minds of an enemy populace, and how to terminate wars and devise exit strategies successfully. None of these problems is new in the history of warfare, of course, but they have achieved particular saliency in the United States because of Korea, Vietnam, and the first Gulf war.

After that war, with an unrepentant Saddam Hussein still on the scene making threatening gestures, fourth and fifth problems—how to counter weapons of mass destruction (WMD) and how to conduct urban combat—came to the fore in the literature. Most recently, with the terrorist destruction of the twin towers of the World Trade Center in New York and our continuing wars of reprisal against the Taliban in Afghanistan and against al Qaeda in that country, Iraq, and elsewhere, sixth and seventh problems—how to neutralize terrorists and how to achieve homeland security—have come to monopolize the pages of the scholarly journals as well as the influential metropolitan dailies, except that now those problems face us and the West in general, not just Israel.

But it is not our purpose here to conduct a review of the thematics of contemporary military literature. All serious students of national defense, plus even casual observers of newspaper front pages and the sound-bites of TV evening news programs, will attest that public analysis of asymmetric war, hearts and minds, exit strategies, urban warfare, WMD, terrorist activities, and homeland security has assailed their eyes and ears *ad nauseam* for the last several years. Unfortunately perhaps, but inevitably, these tired themes are now clichéd and hackneyed in the public consciousness.

The remarkable aspect of these analyses is not their persistency or frequency, however, but rather that, despite their constant presence, the analyses seem so rarely to translate into operationalized success on the battlefield and afterward. The intellectual and scholarly community analyzes problems down to the last quark, but the analysis does not do us much good when payoff time rolls around and we need to apply in war the fruits of our peacetime cogitations.

Observant readers will note that the seven thematic security issues broached above are interrelated to a large degree and that all have coalesced in an incredibly thorny problem complex as the second Gulf War now approaches its uncertain denouement. Tons have been written on asymmetric warfare, yet sectarian militias and masked gunmen armed with rifle-propelled grenades and roadside bombs still fight our modern war machine to a standstill. Skilled behavioral psychologists have delivered the final word on wooing hearts and minds, yet the oppressed whom we come to liberate despise us and cheer over the mangled bodies of our dead soldiers. Political scientists produce impressive studies on winding down wars successfully and extricating ourselves gracefully, yet we must leave large guaranty forces behind indefinitely. Military thinkers devise Solomonic tactics and weapons for conducting urban warfare, but our forces fail to apply them for fear of killing civilians and destroying sacred mosques that the enemy himself does not hesitate to profane. We as a nation fret about WMD in the hands of rogue states, but take decisive action only against the one state that apparently lacked them. We declare a global war against terrorism, yet fail to consolidate our initial victory over the Taliban and al-Qaeda in Afghanistan where it all began. We preach homeland security to the high heavens, but fail to back our preachments with truly serious investment, government reorganization, and citizen sacrifice.

The gap between our supposedly keen analytical solutions to problems of current wars on one hand, and our spotty record in applying those solutions on the other, raises important questions. Is our analysis any good? If it is, does our government pay any attention to it? For example, the so-called Weinberger doctrine counseled, among other things, that no military intervention be undertaken without decisive force. Yet, we conducted Operation IRAQI FREEDOM on a shoestring, succeeding brilliantly in the initial assault against Saddam's organized forces but never being able to muster the sort of widespread, smothering presence that would have snuffed out all significant resistance from the start. Moreover, despite the doctrine's insistence on clear political objectives, it appears that the campaign plan was focused mainly on achieving a quick military victory, with relatively little attention to such politico-strategic concerns as post-conflict consolidation and government reconstitution.

Finally, the most sobering question raised by the continuing gap between theory and application is this: Within the parameters set by political realities, national attitudes, and cultural mind-sets in the United States today, are the seven problems posed by the open-ended war on terrorism in the 21st century even amenable to solution in any kind of decisive, conclusive sense? If not, then how will the United States and its allies defend themselves in the 21st century's war on terrorism, and what should our aims be?

The 22 conference presenters whose views are summarized in the following pages have assembled an impressive body of information and ideas bearing directly on the question of broadening our definition of victory in war to include the coequally valid desideratum of an acceptable peace. This idea is as old as Clausewitz, of course, and we may note further that, since the inauguration of the most recent Clausewitzian renaissance by Michael Howard and Peter Paret in 1976, the nation's political and military leaders have been literally drenched in recollections of the great philosopher's enduring dictum. Yet, in an irony bordering on the surreal, we as a nation continue to celebrate the heroics and drama of battlefield victory as signaling war's triumph, only later waking to disillusionment as the promised political rewards remain tantalizingly beyond reach.

It is perhaps understandable that we fell into such strategic traps in Korea and Vietnam, conflicts arriving on the heels of a world war in which unconditional surrender of enemy arms seemed to be a sufficient goal. But it is far less understandable today, when the Clausewitzian nexus between war and politics is so indelibly engraved on every policymaker's worldview—or so we thought.

In the broad context of demonstrating that wars are won only if the peace is won, as well as suggesting concrete means of capitalizing on this axiomatic truth, our presenters also contribute seminal thought on the continuing wars in Afghanistan and Iraq and in the global antiterrorist war generally. Finally, readers will find precipitating from the collective discussion enlightening cues and insights regarding such macro issues as what Western aims should truly be in a new old kind of war that is global in scope, unlimited in duration, and uncertain in outcome. The 22 presentations, which—it is to be emphasized—represent the views of the speakers and not those of the editor or the Army War College, are commended for reading by all defense theoreticians, policymakers, and practitioners, not only to impart conceptual understanding, but in the ever-renewable hope that they will exercise constructive influence in some modest fashion on the actual conduct of American statecraft.

CONFERENCE PRESENTATIONS AND DISCUSSION

Lieutenant General Jay Garner, Keynote Address.

In Lieutenant General Garner's view, "the American people have never come to realize what terrible shape the Iraqi economy was in nor the desperate plight facing the Iraqi people when the Coalition assumed control" following the U.S. march into Baghdad in January 2003. For example, there was no operational civil communications system and little electricity and potable water. Sixteen of the twenty-three government ministerial headquarters buildings were unusable. Infrastructure had crumbled through long neglect and lack of capitalization. Indigenous police and security forces had evaporated. Accumulated garbage was so voluminous, it required the immediate hiring of 11,000 trash collectors just to avert a threatened public health crisis.

The situation could have become far worse, but four of the most seriously feared problems—Iraqi use of chemical weapons, massive flows of refugees, starvation-level food shortages, and disease epidemics—failed to materialize. Moreover, Coalition interagency planning had been very good, though the requirement to reconstitute the police forces and government ministerial staffs and civil workers presented formidably persistent problems. The organization of the Office of Reconstruction and Humanitarian Assistance was seriously impeded in the early days by wrangling with the Department of Defense (DoD) over selection of the team members who would oversee reconstitution of the Iraqi ministries. Coalition difficulties were further compounded by bureaucratic impediments to timely letting of contracts with the large construction companies that would handle the actual rebuilding. Fortunately, the Iraqi school and university systems have remained functional throughout. The technical skill level among the population at large is more than sufficient for reconstituting and running the country if the insurgency can be contained.

The insurgency that continues today is best thought of as a contest of wills. The United States has a track record of walking away in such situations. A majority of Iraqis are thankful to the United States for removing Saddam Hussein. The Shiites, however, who had most to gain from liberation, will be of little use in rebuilding the country because of their lingering resentment of the massive slaughter of Shiites (perhaps up to 500,000 deaths) that followed the first Gulf war, when Coalition forces pulled out of the country under terms and conditions enabling Saddam's surviving forces to undertake a murderous reprisal. The Kurds in the north, whom we protected, remain a strong support.

Iraq remains an incredibly unpredictable theater, but the Coalition can prevail. Toward that end, the Coalition must do five things: (1) destroy the insurgents; (2) rapidly reestablish the regular army; (3) stimulate the Iraqi economy by cash infusions disseminated directly to families; (4) share oil revenues equitably among all segments of the population; and (5) through a constitution written by Iraqis, establish a carefully contrived mechanism for national governance based on the federal system, with power distributed between a relatively weak central government and a number of empowered constituent territorial governments that reflect the ethnic and religious divisions of the country.

Every Iraqi has an identification card so that fair elections are technically feasible. However, we cannot replicate Western-style democracy in Iraq or other states of the region. Foreign investment and a capitalist economy are pre-conditions for the growth of a democratic culture. The United States and its Western partners need to relieve the Iraqi people of the crushing financial debt incurred by Saddam. Moreover, the United States must have the stomach to see the conflict through, and both political parties must support the effort. Also, we need to trust the Iraqis: "It is better for the Iraqis to do it themselves imperfectly than for us to do it in their behalf perfectly." There are good-news stories from Iraq, but they are crowded off the front pages by the news media's fixation on firefights, roadside bombs, and American casualties. The writing of a constitution in 9 months by fiercely divided Iraqi delegates, for example, was an incredible accomplishment, despite the reservations of Shiite cleric Grand Ayatollah Ali Sistani.

Two encouraging thoughts are worth keeping in mind: (1) the attitude of U.S. servicemembers in Iraq is positively inspiring; and

(2) our soldiers, marines, and sailors will prevail if we as a nation back them up.

PANEL I: *AMERICAN GRAND STRATEGY IN THE GLOBAL WAR ON TERRORISM*

Professor William Wohlforth.

As the lead-off speaker in the first panel, Professor Wohlforth observed that the standard critique of the present U.S. grand strategy as propounded by the Bush administration can be summed up in two words borrowed from critiques of an earlier era: imperial overstretch. According to this critique, the war on terrorism is out of control; the preemption/prevention doctrine is dangerous and counterproductive; unilateralism undermines key alliances and partnerships; dissemination of democracy is beyond our means and wisdom; the doctrine of military dominance is provoking and inflammatory. Taken together, these criticisms coalesce in the judgment that the present U.S. grand strategy reflects a serious overestimation of American power.

A sounder view, however, takes the opposite stance—namely, "that the main flaw of the current U.S. grand strategy is its failure to appreciate U.S. strengths and thus capitalize upon those strengths in fashioning its counter-terror strategy." Based on any scale of power—economic, military, technological, geographical, demographic, institutional—the United States "enjoys dominance on the world stage." Globalization, far from undermining American power as is commonly alleged, actually enhances it—by reinforcing U.S. military-technological dominance, leveraging U.S. influence in the World Bank and International Monetary Fund, increasing U.S. goodwill in developing countries by opening up markets, and enabling the more flexible and better-positioned U.S. economy to reap the benefits of worldwide economic integration. True, globalization in combination with U.S. power and dominance adds to the resentment problem, but on a net basis the positives afforded the United States by the present strategic setting "massively outweigh the negatives."

Given such a strategic setting, three recommendations for revamping U.S. grand strategy follow. First, "emphasize the indirect approach rather than the direct approach in fighting terrorist threats," that is, induce other governments to adopt policies that minimize terrorist threats to the United States. For example, by assisting Pakistan on debt relief, economic sanctions, aid, and a possible textile quota increase, the United States has been afforded unprecedented cooperation from President Musharraf in operations against al-Qaeda. There are a couple of reasons for the indirect approach: it plays to our strengths because they are usually better suited to influencing other governments than to finding and assaulting terrorists; and it is likely to be more effective because local governments are best equipped to police their own territory.

Second, "increase the salience of economic statecraft in the war on terrorism." To return to the Pakistan example, because of opposition from the domestic textile lobby, the U.S. Government reneged in its pledge to increase Pakistan's textile import quota, thereby jeopardizing further cooperation from a leader who desperately needed the increase to bolster his own political standing. The Bush administration should emphasize the connection between unpopular economic policies and U.S. security, so that people come to understand that some domestic economic sacrifices must be made in the war on terrorism.

Third, the United States should "overhaul its approach to multilateral cooperation." The standard critique of the Bush administration in this regard is that it failed to defer to international institutions, thereby alienating allies and partners. This critique is partially correct, but it fails to acknowledge that those not our friends try to use international institutions to constrain the United States in its pursuit of security and national interests.

The correct approach is to adopt a strategy consistent with U.S. power realities. There are two strategies that play to American strengths. Policy 1 is simply to resign from the national order, breaking free and reaping the advantages that total autonomy would bring. The downside is that we would forfeit what are often the very real benefits of association with other states—cooperation, assistance, and enhanced legitimacy. The other feasible strategy, Policy 2, is to participate vigorously in international institutions and seek to shape the world order toward our own ends. The downside is having to submit to constraints that run counter to our interests when, regardless of our power, we cannot carry the day in the governing forums.

The key criticism of the Bush administration strategy is that it has followed a waffling middle course between Policies 1 and 2, suffering the disadvantages of both while harvesting the advantages of neither. Under what we might call Policy 3, it has exempted itself from institutional restraints at times, but later gone back to the institution with hat in hand, begging for help in exchange for willingness to be constrained. The result is the worst of both worlds—the United States increases the apparent importance of the international institutions even as it does nothing to reshape them in ways that will benefit us.

Professor Wohlforth states that, while he lacks the research to back up a recommendation for either Policy 1 or Policy 2, he certainly recommends that "we move away from Policy 3."

Dr. Robin Dorff.

Dr. Dorff takes the position "that a genuinely comprehensive approach to promoting legitimate national governance within the world's community of nations is an appropriate and necessary U.S. grand strategy for achieving greater national and international security."

The attacks on the World Trade Center and the Pentagon, followed by our subsequent military operations in Afghanistan and Iraq, have heightened awareness of and interest in both terrorism and so-called failed and failing states. This awareness has helped to broaden the security debate and forced a somewhat reluctant policy community in the United States to recognize the changing face of security. Whether we choose to call it nation-building, post-conflict reconstruction, or something else, the point is simply that we can no longer view security in strictly military terms (if that were ever the case), and we must accordingly engage the full array of national and international elements of power in pursuit of security. Economic, political/diplomatic, and informational tools must be coordinated and focused on the task of building an international assemblage of nations whose governments bask in the sunlight of legitimacy, human rights, and popular support. This task must be accomplished in an increasingly interdependent world and amidst an expanding number of global challenges. In short, "what the United States needs is a grand strategy that focuses on the global deficit of good governance."

Professor Robert Lieber.

The threat environment is defined primarily in terms of the following three factors: (1) the implications of September 11, 2001; (2) the inherent weaknesses of international institutions; and (3) the necessity for U.S. self-help and a risk-acceptant posture. The day we have come to call 9/11 marked a watershed in the nation's threat environment, one of those few genuinely apocalyptic moments such as the British torching of Washington during the War of 1812 and the destruction of the U.S. Pacific fleet at Pearl Harbor in 1941. In 9/11, the profound threat from WMD, terrorism, and radical Islamic movements coalesced in a single convulsive act. The weakness, paralysis, and ineffectuality of such multilateral responses to the threat as the United Nations (UN) and European Union might muster mandate an American policy of "self-reliance, independence, and even unilateralism."

The long-familiar strategy of containment is ill-adapted to provide security in the face of the new threat because the enemy is at once so amorphous and fragmented that American power cannot be applied in a deterrent role. Moreover, if deterrence fails, the consequences could be horrific. Thus a more aggressive strategy is indicated, allowing for preemptive and preventive attacks by the United States to forestall intended terrorist attacks. Like the Cold War, the war on terror is likely to be long and costly, but it is unavoidable.

With regard to Iraq, "debate should focus not upon whether the war was legal (it was), but rather upon its wisdom." In this regard, Saddam did constitute a strategic threat because of his past demonstrated willingness to employ WMD and because of his attacks against his neighbors, which could have destabilized the region. The dangers of not acting outweighed the dangers of acting, even in the presence of imperfect information. There are obvious difficult postwar issues to be dealt with, but they are resolvable if America stays the course. The Bush administration's response is necessary in view of the genuine long-term threat.

PANEL II: *FISCAL AND POLITICAL SUSTAINABILITY*

Dr. Alice Rivlin.

All of the actions being discussed in the present conference—coping with terror and insurgency, peacekeeping, nation-building, democratization—depend on resources as well as strategy and leadership. They depend on the strength of the American economy and the willingness of the public and their representatives to dedicate resources sufficient to implement them.

The outlook for the U.S. economy is favorable, but the prospect for the federal budget is extremely worrisome. If current policies continue and the tax cuts are extended, the federal budget is likely to run large sustained deficits for the next decade, which will constrain military as well as domestic spending. After that, the situation gets worse as spending for retirement programs rapidly escalates. We are a wealthy, highly productive country. We can afford both a strong defense and adequate domestic public services. But "we can't do it on borrowed money."

Looked at objectively, the current statistics about the economy give a mostly good-news story. Slow job growth, however, is very serious if you are out of work or worried that you might be. Optimism is justified on this score, first because of the economy's strong job creation in the past decade, and second because of the impressive resilience of the economy in the face of such shocks as the stock market crash, 9/11, wars in Afghanistan and Iraq, and the wave of corporate scandals.

There is no question that the federal deficit is high—about half a trillion dollars or about 4.5 percent of the Gross Domestic Product. But this deficit is not a problem now because the economy is operating below capacity, and inflation is tame. Depending on which level of economic growth one predicts for the balance of this decade, near-term deficits could vanish or they could worsen. But after 2010, in Dr. Rivlin's view, "rising demographic pressure will overwhelm the budget, no matter what growth rates are."

Deficits do matter, primarily because they eventually entail higher interest rates which can damage the economy, and because they transfer debt burden to future generations. Moreover, there is the additional worrisome uncertainty surrounding the huge amount of our public debt held by foreigners. Should they ever decide to dump their dollar-denominated investments, a financial crisis could ensue.

Recently, a Brookings team conducted comparative studies of how the federal budget could be balanced in 10 years. Despite drastic (even draconian) spending cuts and despite shifting heavy burdens to state and local governments, not one of the three approaches adduced could balance the budget by 2014 without substantial additional infusions of new revenue.

It thus becomes extremely important for those involved in discussions on U.S. grand strategy to think constructively about the wider context within which decisions on national security expenditures will be made. We are "in a deep hole, and it will get deeper soon." We shall have to make difficult choices on both the spending side and the revenue side of the budget. Success will require a strong bipartisan consensus on the objective of getting to budgetary balance and on restoring budget decision rules requiring hard choices. Toward these ends, the study team made four proposals: (1) restoring caps on discretionary spending; (2) restoring pay-as-you-go rules for both entitlements and taxes—meaning that new entitlements and tax cuts have to be paid for; (3) imposing strict definitions of emergency exceptions; and (4) prohibiting the passage of sunset laws (those laws stipulating termination dates for government programs absent specific congressional reauthorization) just to make cost look smaller.

Dr. Eric Labs.

Long-term cost projections of resources for defense are trending higher, even overlooking the huge spike occasioned by the FY03 Omnibus and Supplemental Appropriations Bills and the FY04 Supplemental, which together were necessitated by the wars in Afghanistan and Iraq. Resources for defense include the categories Operations and Support (for the salaries of DoD's military, civilian, and contractor employees, their benefits, the operating costs of DoD's equipment, and the costs to operate and maintain defense facilities) and Investment (development and procurement of DoD's weapon systems). DoD's total obligational authority (TOA) grew rapidly from the early to mid-1980s, reaching a peak of $420 billion in 1985 at the height of the Reagan defense buildup. Reflecting the end of the Cold War, TOA then generally declined during the late 1980s and into the 1990s, reaching its lowest point in 1997 at about $238 billion.

DoD's obligational authority then began to rise again, reaching $304 billion by 2000 and then growing rapidly as U.S. forces became engaged in operations in Afghanistan and Iraq. In 2003, DoD's TOA reached $449 billion, including $74 billion in supplemental funding. The 2004 Future Years Defense Plan (FYDP) anticipated that defense resources (excluding supplementals) would rise from $383 billion in 2004 to $439 billion in 2009. If supplemental appropriations are included, obligations in 2004 will be some $450 billion, about the same amount that Congress appropriated for DoD in 2003.

If the program in the 2004 FYDP is carried out as currently envisioned, the demand for defense resources, exclusive of contingency needs, would continue to increase through 2022. They would average $458 billion per year between 2010 and 2022—or about $75 billion more than the 2004 request, and about $8 billion more than the funding DoD received in 2004. The projection for 2010-22 uses DoD's cost estimates for its planned programs, and, consistent with the 2004 FYDP, excludes costs for continued operations in Afghanistan, Iraq, and other activities conducted in the global war on terrorism.

Dr. Labs notes that various factors "could push the costs of current plans even higher": for example, costs for weapons programs might grow as they have since the Vietnam War; operating costs for major equipment might increase as they have over the past 2 decades; and the United States might continue to conduct military operations overseas as part of the continuing global war on terrorism. Should these assumptions prove correct, and there is a high risk that they will, then resource demands would instead average about $473 billion a year through 2009 (12 percent higher) and about $533 billion between 2010 and 2022 (16 percent higher).

The Operation and Support basket accounts for about two-thirds of the DoD budget. Expenditures are projected to rise from $236 billion in 2004, to $254 billion by 2009, to $292 billion by 2022, with health care costs (expected to nearly double between 2004 and 2022) and real pay raises accounting for the increase.

The Investment basket, constituting about one-third of the DoD budget, is projected to grow from $137 billion in 2004 to $171 billion by 2009 to $186 billion by 2013, after which it would decline. Over the 2010-22 period, assuming that weapons costs do not grow as they have historically, investment demands would average about $175 billion a year. These figures exceed the January 2003 projections of the Congressional Budget Office (CBO), the excess being accounted for by increases in procurement spending on the Army Future Combat System ($9 billion a year), defense agency funding increases ($8 billion a year), a new transformation-related research, development, test, and evaluation (RDT&E) account ($5 billion a year), and Air Force and Navy investment increases (about $6 billion and $1 billion a year, respectively).

The CBO projects that if the Army's current plans are carried out, by 2012 the Army's investment spending will exceed the $34 billion spent in 1985, and by 2014 it will reach a peak of $42 billion. With the Future Combat System (FCS) driving the Army's high cost projections, the CBO continues to project increased demands for Army investment over the next decade despite a number of retrenchments. These include canceling the *Commanche* reconnaissance helicopter program; terminating the planned upgrade programs for the Army's current fleet of ground combat vehicles, including *Abrams* tanks and *Bradley* fighting vehicles; delaying the start of FCS procurement; and transferring funding for destruction of chemical munitions out of the Army's budget.

Dr. Michael O'Hanlon.

Budgetarily, the United States had better prepare itself to make some very tough choices regarding the size and weaponry of its forces. Whoever wins the November 2004 election will have to get serious. The Army faces the gravest problems and needs to move now. If it waits, it will face a crisis without a quick solution.

The Army, and possibly the Marines, needs an immediate increase in its active-duty troop level. The decision is badly overdue, with the problems worsening daily (the decision to increase should have been made in mid-2003 after it became obvious that the post-Saddam stabilization mission in Iraq would be long and difficult).

There is no definitive method for determining the appropriate size for the Army because "it is impossible to figure exactly how large a rotation base would be needed to continue the Iraq mission over a period of years, while avoiding unacceptable strain on the all-volunteer force that could drive large numbers of soldiers out of the military." But we do know that today's policies, if continued for long within today's parameters, carry with them the very real risk of breaking the all-volunteer force.

"Logic and a basic sense of fairness" suggest that we generally should not send active-duty troops back to Iraq after only a short respite at home between successive deployments. One year in Iraq, one year home, and then another year in Iraq constitute an extremely demanding pattern. Yet, that is exactly what the Army will soon need to do with some units. Moreover, the Chief of the Army Reserve has argued convincingly that reservists should not have to be involuntarily activated for more than a 9-to-12 month period once every 5-to-6 years.

One simplified but illuminating way to approach the problem of estimating the necessary increase in troop strength is to figure how many new units would be needed to carry out the Army's share of an entire rotation in all overseas missions—notably, Iraq, Afghanistan, and Korea—in 2006 (it would have been much safer to generate the respite in 2005, but given the Pentagon's reluctance to espouse such a policy, it is now too late for that year). Providing a break for existing units in 2006 is now the absolutely minimal objective for relieving strain on the active and reserve forces in the coming years.

Roughly speaking, the Army and Marines may have to provide 85,000-110,000 ground troops for these missions in 2006, assuming a requirement for 50,000-75,000 in Iraq, 10,000 in Afghanistan, and 25,000 in Korea. Perhaps on the order of 15,000 of these can be provided out of the combat brigades of the Army National Guard, leaving a need for 70,000-95,000 new soldiers and marines. Of that number, 10,000 should be generated from existing and planned Pentagon initiatives to privatize certain current military positions.

Another 15,000 (plus or minus) active troops may be generated internally from force rebalancing measures by which high-demand units are augmented with soldiers from lower-demand units such as artillery. The resulting arithmetic indicates that about 45,000-70,000 new troops would have to be added to present end strength.

Secretary of Defense Donald Rumsfeld already is planning to use emergency powers to increase the size of the active ground troops by some 20,000, so that a net addition of 25,000-50,000 would be required according to the foregoing rough calculation. Some of the increase might need to go to the Special Forces. But given the nature of Special Forces—elite, older, very highly specialized and trained—and considering the pressing need for additional combat and support units, most of the new ground troops should go to conventional Army formations.

The increases in troop levels and the enormous expenses of operations in Iraq and Afghanistan will absolutely mandate economies, retrenchments, and scale-downs in certain other sectors of the total defense program. The likely candidates for reduction, according to Dr. O'Hanlon, are the following.

Missile Defense. By 2005, missile defense spending will be at the rate of about $10 billion a year, a figure expected to continue for the rest of the decade. This amount is roughly double the annual amount spent in 1991 during the elder Bush's administration.

Despite the expense and immaturity of the technology, deploying an interim missile defense capability in the coming months makes sense. The missile threat is real. Strategically speaking, the United States has absolutely no missile defense capability now. Even an imperfect system is better than the status quo. Rudimentary fortifications "based on throwing up barricades is preferable to leaving a city naked before invaders."

Since missile defense is not now the country's top security priority, however, some sense of perspective is in order. Even as it continues to deploy an interim long-range missile defense capability, improve shorter-range missile defense systems, and work on technologies for better future systems, the United States should scale back its missile defense plans, perhaps to a level between that of the late Clinton years, when resources were devoted primarily to research and development, and the current Bush outlays. That would allow for deployment of perhaps a two-tier strategic defense system, designed to intercept a few dozen warheads or less, as well as theater missile defenses and continued intensive R&D efforts.

Additionally, several specific economies should be considered such as reducing the $1 billion intended in 2005 for the Navy's *Aegis* destroyer platform for a long-range missile defense system; limiting any deployment of the main midcourse missile defense system to 50-100 interceptors, enough to deal with the more plausible attack by North Korea; canceling the acquisition of certain ancillary technologies such as low-orbit tracking satellites; and stretching out the overall program.

Air Force. Open to serious doubt is the Air Force's sustainment of two huge simultaneous main combat weapon programs—the stealthy air-supremacy/ground-attack F/A-22 and (in conjunction with the Navy and Marine Corps) the stealthy ground-attack Joint Strike Fighter (JSF). This is particularly true given that recent Air Force operations have succeeded with stunningly few losses; that improved electronics, sensors, targeting techniques, and communications have greatly increased the precision and thus the efficiency of existing aircraft; that refurbishing present aircraft or replacing them with newly produced versions (e.g., F-15s and F-16s) would produce enormous savings; and that the improved performance of unmanned aerial vehicles throughout the next decade promises that human pilots will be supplanted in some roles.

There is thus a strong presumption that the F/A-22 *Raptor* is not needed and that refurbished or new F-15s would suffice for years to come. "Given the maturity of the *Raptor* program," however, there is a case for its continuance on schedule. If that occurs, the JSF program should be delayed by at least 2 or 3 years, and the total purchase should be reduced from 3,000 to 1,000 aircraft. The net savings from these and associated changes would be about $50 billion, with over half accruing over the next 10 years, mostly for the Air Force.

There are also sensible savings to be had in the Air Force (and Navy) budgets with regard to nuclear weapons. Russia and the United States are committed to reducing their numbers of operational strategic warheads to between 1,700 and 2,200 by 2012, with no other stipulations. Therefore, "the United States should do it sooner rather than later, and in a manner designed to save money." With regard to platforms, we could retire 200 *Minuteman* missiles and convert six (instead of the planned four) *Trident* submarines to an all-conventional role, thus according to the CBO saving $1 billion per year.

Army. The Army's prime candidate for economies is the Future Combat System (FCS), the next-generation replacement for the main battle tank as well as other elements of the current heavy divisions. It is conceived to weigh far less than an *Abrams* tank and therefore to be more deployable (fitting on a C-130 aircraft) and fuel-efficient, depending on information networking rather than heavy armor for its survivability. A production decision is planned for 2008, with initial operational capability by 2010 and one-third of the Army having fielded the FCS by 2020. The total cost of these efforts is estimated to be $92 billion.

The concept is reasonable but "seems rushed," given that the *Stryker* program is to form a half dozen medium-weight brigades and is itself in the process of implementation, with a sizable legacy heavy force still in existence. Slowing the main objectives of the FCS program by as much as a decade seems reasonable. That would space the *Stryker* and FCS generations 15 years apart, still a rapid pace for modernization, thus saving about $3 billion or more a year during the 10-year interim.

Navy and Marine Corps. Apart from the *Joint Strike* Fighter, *Aegis* destroyer, and *Trident* submarine issues broached earlier, the most expensive program of dubious value is the V-22 *Osprey*, a tilt-rotor aircraft long plagued by fatal accidents during trials. Its expected survivability in combat is only 10 to 20 percent better than a helicopter's, and its crash-avoidance flight profile will probably not be feasible in combat conditions. Moreover, helicopters will still be necessary to carry the heaviest equipment during aerial assaults. Viewed as a niche-mission aircraft, and one still under development at that, the V-22 lends itself to a modest purchase of only a few dozen. The net savings would be about $10 billion over the life of the program.

The Navy plans to buy 16 of the new DDX land-attack destroyers. But it could save nearly $1 billion a year over the next decade by obtaining only half that many, while basing four of those overseas or at least keeping four overseas for 2-year stretches at a time. Similarly, additional savings could safely be achieved by buying fewer *Virginia* class submarines, compensating by forward-stationing up to a half dozen at such bases as Guam.

PANEL III: *INSURGENCY AND TERRORISM*

Dr. Steven Metz.

In taking up the topic of insurgency, Dr. Metz begins with definition: Insurgency is a strategy in which the weak use various forms of protracted asymmetric violence, psychological conflict, and a radical counterideology to alter the balance of power in their favor.

Insurgency first pursues (through necessity) a protracted alteration of the power balance and only then an imposition of will on the target group or population. It aims to mobilize support for the insurgents' cause at the same time it is eroding support for the existing power. It can target the balance of power within a nation (the most common practice during the mid-to-late 20th century), within an entire region, or even globally. It can seek outright victory or something less. Finally, insurgency "is usually a long-shot strategy adopted out of desperation."

The contemporary U.S.-dominated world order seethes with resentment and hatred in many quarters, but those opposed cannot alter the existing world through conventional military means. Hence they are seeking to alter that order through unconventional means, i.e., through insurgency. Insurgency has always existed because there have always been weak groups that wanted to supplant strong ones, but historically such challenges to the existing order have ebbed and flowed in strategic significance. Insurgency today has entered a period of enormous strategic import, one bold enough to challenge the most powerful nation on earth along with its Western allies.

The 20th century was a "golden age for insurgency" because the breakup of the colonial empires left in its wake weak, unstable states seemingly ripe for the plucking, while the Cold War provided generous external sponsors of contending groups. Today, however, insurgency is different. Because of such technological advances as instant international news dissemination, television, cell phones, email, and the worldwide web, insurgents have the advantages of greater cohesiveness, unity of effort, and intra-group communication. Moreover, they can carry their message to like-minded groups throughout the world, seeking and often gaining support. Since they no longer can acquire the large state sponsors characteristic of the old bipolar world, they must generate their own revenue (though they can still at times obtain under-the-table financial backing from sympathetic or fearful states).

But unlike the communist and nationalist ideologies of the Cold War era, which provided both a critique of existing orders and a countervision, radical Islam "issues a critique but can offer no acceptable alternative." Thus insurgency born of Islam is entirely negative, and the flaws of its radical alternatives are laid bare before the world by the same modern communications that carry the insurgency's message.

The United States is in the process of working out its decision as to whether to pursue a strategy of victory or a strategy of management. It has declared a strategy of victory, but assumed it could successfully pursue such a strategy without a traumatic mobilization of means. It is discovering, however, that a successful strategy of victory would entail undertaking a second track of transformation, i.e., such difficult and painful measures as a truly cross-governmental effort (since counterinsurgency is 80 percent political, economic, and psychological); a professional core of regional security and development advisors; integrated national intelligence, strategy, and operational planning; culturally astute military units capable of protracted operations; effects-based planning and execution to eradicate the insurgency threat in all dimensions; and a sustained effort based on the full integration of contractors.

A strategy of management, on the other hand, would "entail acknowledgment that so long as there are haves and have nots in the world, there will always be insurgencies." The United States as the richest, most powerful, most resented, and secretly most envied nation in the world will never achieve absolute victory over insurgents in the sense of putting an end to their efforts. Faced with these realities, the strategy of management would reduce the scope, aggressiveness, and intensity of its overseas military actions against insurgents to a sustainable level, confining its major overseas initiatives to the political, diplomatic, and psychological realms. In this country, institutionalizing homeland security for the long haul would be the main concern.

However, with the American declaratory policy of victory combined with a failure thus far to make the necessary material and organizational investment, we have a "means-ends mismatch," with a level of effort more appropriate for a strategy of management.

With respect specifically to the Iraq insurgency, three points are to be noted. First, the insurgency was bottled up at first, emerging only gradually as the Coalition began efforts to turn the government over to the Iraqis. There has thus been a 10-month learning contest between the insurgents and the Coalition. Second, following the American tradition, we have done better with the military dimension than with the political, social, and psychological. Insurgencies that morph into a nationalistic struggle increase their chances of success. That could happen in Iraq. Third, the key determinants of the future will be the degree of legitimacy of the new Iraqi government, the effectiveness of the security forces, and whether the insurgent factions can coalesce in a united front.

In considering insurgency as a global phenomenon, Dr. Metz makes four observations. First, in the face of the U.S.-dominated and enforced global order, insurgency is now a major mode of conflict because those opposed to that order have no other means to alter it. We are in the midst of the first global insurgency, and it will be the dominant motif of U.S. grand strategy in the coming decade. Second, at one level the antagonists are the United States and its allies against radical Islam. But at a deeper level, the antagonists are the people and entities who have adapted to globalization against those who continue to oppose it. Third, the global insurgency will be extremely hard to eradicate, but it is built on a narrowly negative ideology which we have all the necessary conceptual tools to counter. Fourth, we need a psychologically astute grand strategy, one that can placate anger and frustration on a global scale, one that can give the disenfranchised a stake in and enable them to identify with the system.

The vital questions:

- Is it possible to ameliorate the widespread discontent with the American-dominated world order? This is an extremely difficult question. The liberal perspective holds that the root causes are manipulatable and thus subject to ameliorative reforms. The conservatives hold that the causes are cultural and immutable, and in any event are not remediable at an affordable cost.
- Does a strategy of victory over global insurgency make sense given the great cost and risk? Thus far, despite having declared a strategy of victory, the United States has been loath to undertake the sort of fundamental cross-governmental transformation necessary to implement it. This means-ends mismatch is a "recipe for disaster."

Professor Ian Beckett.

To place the insurgency in Iraq in an adequate historical perspective would suggest taking note of the Arab revolts against the British in Iraq in 1920 and in Palestine between 1936 and 1939; the U.S. Marines' experience in Nicaragua, Haiti, and the Dominican Republic in the 1920s and early 1930s, and in Vietnam with the Combined Action Program in the late 1960s; the successful Jewish insurgency in Palestine between 1945 and 1947 to force out the British; the largely secular National Liberation Front's (FLN) ultimate success against the French in Algeria between 1956 and 1962; the Arab insurgency against the British in South Arabia, specifically, the actions leading to the British abandonment of Aden in the late 1960s and their successful Dhofar campaign in Oman during the period 1965-75; the British experience in Northern Ireland since 1969; and the Palestinian tactics against the Israelis today, as well as the anti-Israeli attacks by Hezbollah in Lebanon in recent years.

Islamic fundamentalism, "which might be regarded more as an ideology than an expressly religious conviction," has emerged as a new imperative behind insurgency ranging from the struggle against the Soviets in Afghanistan between 1979 and 1989 to the continuing conflicts in the Philippines, Indonesia, Algeria, Sudan, Kashmir, and

Chad. Of course, some of these conflicts may also be characterized in other terms such as ethnic or separatist insurgencies.

Notwithstanding new motivational impulses, however, and the fact that insurgents are increasingly better armed, perhaps more fanatical, and in some cases better attuned to the information revolution than in the past, it is still the case that much remains the same in terms of the basic requirements for successful insurgency. Insurgency remains a highly political act arising from some sense of grievance, or upon the exploitation and manipulation of grievance. An insurgent leadership is still likely to be better motivated than the rank and file. Insurgency will still be the recourse of those initially weaker than their opponents and, though perhaps less protracted than in the past, may still largely depend for its ultimate success on substantial external support. Above all, however, insurgency is still invariably a competition for governmental control and in perceptions of legitimacy.

One fundamental question is whether what is being faced in Iraq is insurgency or terrorism or perhaps merely a traditional form of guerrilla warfare or resistance. Prior to the 20th century, guerrilla warfare was understood as a purely military form of conflict, with classic hit-and-run tactics employed by indigenous groups in opposition to foreign or colonial occupation. Rarely did such guerrillas display any wider comprehension of the potential of irregular modes of conflict, and it was only in the 1930s and 1940s that guerrilla warfare became truly revolutionary in both intent and practice. Thus, dissident groups that were initially in a minority and weaker than the authorities would seek power through a combination of subversion, propaganda, and military action.

More properly, therefore, modern revolutionary guerrilla warfare increasingly was termed insurgency, with guerrilla tactics being employed strategically to achieve a particular political and/or ideological end. The transition from guerrilla warfare to insurgency does not depend, therefore, upon the size of any particular group, but upon the intention to bring about fundamental political change through a political-military strategy of organized coercion and subversion, and, usually, also the attempt to mobilize a mass political base. While insurgents might routinely employ terror or intimidation in tactical terms, they have rarely done so at the strategic level. Consequently, it can perhaps be argued that terrorist groups, even if motivated by an ideology similar to that of insurgent groups, have tended to employ terrorism indiscriminately and as political symbolism without the same intention of taking over the state apparatus themselves.

The absence of a real insurgency in Iraq, however, does not make the situation less dangerous, and whatever the motivation of any particular group, the aim is clearly to sow divisions both between the Iraqis and the Coalition and between Iraqis themselves, and to raise the costs of the U.S. and Western presence. There can be little expectation on the part of these various groups that they are capable of challenging the Coalition's military superiority.

The basic patterns of insurgency have not materially changed, and there are similarities between the emerging situation in Iraq and some earlier insurgencies in the Middle East. It follows that the essentials of counterinsurgency have also remained fairly constant and that the kind of basic requirements for success that can be identified in campaigns since 1945 still hold good. These requirements are, first, a recognition of the need for a political rather than a purely military response to insurgency; second, a need for coordination of the civil and military response; third, a need for the coordination of intelligence; fourth, a need to separate insurgents from the population; fifth, a need for the appropriate use of military force, which generally means the minimum necessary in any given situation; and, last, the need to implement long-term reform to address the grievances that led to support for the insurgency in the first place. These principles are as applicable to Iraq as they once were to the British mandate in Palestine, Aden, Algeria, and Oman. Where they were not adhered to, as in Palestine, Aden, and Algeria, counterinsurgency failed; where they were, as in the Dhofar, counterinsurgency succeeded.

Mr. Steven Simon.

The basic question is: "What happened to al-Qaeda?" Views differ as to whether it still exists in Afghanistan, and, even if it does, whether it is still an effective force. Most likely, it remains what it always was, a "unique hub-and-wheel organization" with the brains and money at the center loosely connected to independent elements operating along the rim, inspired and united by the charismatic leader Osama bin Laden at the center. Since our forces went into Afghanistan, the center is now a floating hub, embodied in a leadership that moves around to maintain safety in a succession of secret locations. Thus the amorphous al-Qaeda organization "resembles one of Salvador Dali's melting wheels, an apt characterization in this increasingly surreal struggle."

The al-Qaeda emblem provides a very large tent for the driven, the outcast, the malcontents, the disaffected young Muslims of the world. It asks, are you an anti-colonial activist? It answers, join al-Qaeda. Or are you bereft of prestige and identity? Is your religion under attack? Is your country being robbed of its wealth by the wicked West? Do you hate your government because it has abandoned the true faith? Regardless of grievance, the answer in all such cases is to join al-Qaeda. It thus casts its agenda in protean terms, and its occasional spectacular successes, as in the destruction of the World Trade Center buildings in New York, are hugely inspiring and catalytic to young men and women of the Middle East who can find no other recourse. In this regard, an authoritative reference is Professor Alan Richards' superb monograph titled *Socio-Economic Roots of Radicalism? Towards Explaining the Appeal of Islamic Radicals*, published by the Strategic Studies Institute in 2003.

The grand ayatollahs, who appeal to the higher instincts of Islam in counseling a moderate course of restraint and reconciliation, find their voices ignored, while those of the firebrand mullahs, who cry jihad and incite their followers to insurrection, draw rapt, responsive crowds. In responding to what can only be called a global Islamic insurgency, American rhetoric frequently goes amiss. President Bush speaks of cutting off the head of the snake, while Central Intelligence Agency (CIA) Director George Tenent talks of breaking the back of the enemy. Such analogies betray a naïve judgment that an amorphous, widespread organization like al-Qaeda can be exterminated by a single devastating stroke at some presumed focal point of vulnerability.

Similarly, U.S. attempts to influence Muslim populations are frequently beset by counterproductive messages. For example, an American propaganda video shown in Indonesia depicted a female

Muslim in the United States jogging in shorts, presumably an effort to demonstrate the integration of Muslims into American life. But such a scene would be anathema in most parts of the Muslim world. Influence operations to win hearts and minds are very difficult and tricky. We don't do them well. Al-Qaeda does them far better, often successfully portraying generous, constructive U.S. actions in the Middle East simply as ruses intended to conceal our hostile intentions.

Mr. Simon finds three elements of the global war on terrorism as being especially important: (1) democratization, (2) alliances, and (3) homeland security. Spreading democracy to failed states, along with all the economic and social benefits that we associate with transparency, accountability, and consequent capital investment, is the only credible long-term solution to what ails the globe's huge islands of poverty and discontent. Alliances are essential in combating terrorism, not because they are necessary to encourage direct attacks on terrorists—no country wants them on its soil—but rather for the broad-gauge programs designed to win hearts and minds and stimulate economic development in states that breed terrorism. The job is simply too big for America alone. Finally, we need to get truly serious about homeland security. There is an apocalyptic bent in terrorist insurgency that leads members toward WMD. We must get our house in order because otherwise something unspeakably dreadful could happen.

Mr. William Kristol, Banquet Address.

Mr. Kristol began with the declaration that September 11, 2001, marked a historic juncture in U.S. statecraft and the American sense of security. "Things will *never* be the same." We will never return to the innocent, halcyon days of the 1990s.

Three overriding concerns are reflected in President Bush's foreign policy: terrorism, WMD, and dictatorial regimes. Together, in an intertwined way, they constitute the gravest threat to American security that we have faced in living memory. In reality, the President is focused on terror, not terrorism, and this is proper for the American people will never consent to living out their lives in constant fear of deadly attack from without. Given these three overriding concerns, the President looks to the Middle East as the source of the problem. The President is saying that we can't go on the old way. We have to change the Middle East. We have to neutralize the sowers of terror; we have to find and eliminate WMD and prevent their reappearance; we have to facilitate the emergence of democratic regimes, for it is these that will maintain peace with their neighbors and redress the sense of privation and alienation that afflicts their people.

With regard to concern that the President failed to mount his response to domestic terror with greater alacrity and in a more aggressive manner, Mr. Kristol noted that it took a while for the administration to digest the situation and devise a proper response. The proper response involved very big decisions.

In Iraq, there are many measures we should have taken and failed to take. The situation is probably manageable, but we will have to send more troops. President Bush will send them, even in an election year, because he knows it is right and necessary.

Despite concern that Senator Kerry has not taken a firmer, more resolute stand in support of the American war effort in Iraq, a case can be made that if he wins the presidential election, he will act responsibly, and things won't change much. There is more underlying bipartisanship regarding the war than we realize, but it is obscured by "obsessive electioneering."

The next big threat to American security, in Mr. Kristol's view, will come from China if that nation retains its antidemocratic character.

PANEL IV: *PEACEKEEPING, NATION-BUILDING, AND STABILIZATION*

Dr. Conrad Crane.

In the post-conflict posture following decisive military operations against a target state, we normally envision a transitional reconstruction period during which peace and a viable state are established. The reconstruction period is divided into four distinct phases: I, Provide Security; II, Stabilize; III, Build Institutions; and IV, Handover/Redeploy. This scheme appears perfectly reasonable from a conceptual point of view. Given the compressed timetable of recent rapid deployment operations, however, with the purely military victory often occurring within weeks, days, or even less, it is imperative that reconstruction planning begin earlier than in the past and that reconstruction efforts themselves be initiated during the period of decisive military operations. In fact, for conceptual clarity and a proper appreciation of relative priorities, we should dispense entirely with a Phase IV as such, instead conflating it with a newly designed Phase III, Build Institutions/Handover/Redeploy. In such a conflation, reconstruction tasks might have to be reshuffled somewhat among the three remaining phases, depending on whether the tasks are judged to be critical, essential, or merely important. Such a scheme would lend doctrinal sanction to the battlefield lesson that, with the rapidity of recent stability operations, certain reconstruction efforts must begin "even while the fighting is still in progress."

In sum, reconstruction embraces three vital objectives that must be met generally in the order shown: (1) achieve security within the target state; (2) stabilize local commerce, government, and services so that the populace can resume the rudiments of daily life; and, (3) rebuild the economy and the institutions of government while handing over the reins of authority to appropriate civilian institutions or indigenous officials, and redeploying forces as early as feasible. In the past, no part of post-conflict operations has been more problematic for American military forces than the handover to civilian agencies, which ideally should precede the handover to indigenous authorities. More specifically, to limit the potential for regional backlash from a perceived prolonged military occupation, it is highly desirable that U.S. civilian and international organizations assume coalition military responsibilities as soon as possible. Unfortunately, the reality usually seems to be that the handover is directly to the local government. Clearly, successful reconstruction is as important as winning the shooting war, but within reconstruction, a successful handover/redeployment is every bit as important as building institutions.

State stabilization and reconstruction are missions the American military would prefer not to take on. It does not have the proper force structure or doctrine to perform such missions, and its warrior mentality and culture are seen to be at odds with the requirement for winning over a suspicious or hostile populace. Yet, of all national and international agencies, the American military is the best for such missions when properly motivated and resourced. The Army as the lead ground force will bear the brunt of such efforts. The superb warfighting force that produced victories in Operations DESERT STORM and IRAQI FREEDOM generated a professional mindset antithetical to the roles of counterinsurgency, peacekeeping, post-conflict stabilization, and nation-building. But since these are precisely the roles the Army will most likely be called upon to perform in the global war on terrorism, it becomes the task of Army leaders to shape a force that can win the peace as well as win the war, and to do both willingly.

Ms. Bathsheba Crocker.

We need to think analytically and realistically about the U.S. capacity to do post-conflict reconstruction work. Some question "whether we really want to get better at it because, if we do get better, we'll be called upon to do more of it." The foregoing reasoning is not sound, however, because whether we like to do post-conflict reconstructions or not, we've done or tried to do about one every other year since 1990. Moreover, in view of the ongoing global war on terrorism, Iraq will not be the last. In fact, with the large commitments in Afghanistan and Iraq, not to mention continuing missions in Kosovo, Bosnia, Sierra Leone, Congo, and Haiti, as well as upcoming ones like Sudan, such interventions are overwhelming our national security apparatus. Unless we improve the capacities of the entire U.S. government (by institutionalizing and adequately resourcing functional entities specifically designed to undertake the various aspects of post-conflict reconstruction), then the overwhelming burden will continue to fall on the U.S. military because the "need won't go away, and they will continue to be the only game in town."

The persistency of the problem has unleashed a frenzy of study groups, task forces, committees, and commissions, inside and outside government, all looking for solutions. Some consensus seems to be emerging, with such deficiencies as the following having been identified: (1) lack of strategic planning and coordination capacity on the civilian side; (2) lack of home office and leaders on the civilian side, including lack of a lead civilian agency; (3) no consistent and flexible funding stream; (4) no standby capacity on the civilian side, e.g., legal teams, civilian police, civil administration/elections/constitution/rule of law experts; (5) lack of a targeted training capacity; (6) inadequate security apparatus for doing stabilization/constabulary/public safety work; and (7) lagging civilian/military coordination. Legislation and other efforts would create some new authorities and offices, set up standing coordination mechanisms, create new sources of funding, generate ready rosters of relevant civilian experts, and establish training centers. Had these proposed reforms been in place before our interventions in Afghanistan and Iraq, however, it is doubtful whether we could have avoided some or any of the mistakes and problems occurring in those countries.

Attention shifted from Afghanistan to Iraq fairly quickly, and it has not shifted back, despite some very real concerns. Our efforts in Afghanistan have been called "nation-building lite" or "on the cheap." There are 13,000 U.S. troops plus 6,000 international troops in the country, equating to less than one peacekeeper per 1,000 Afghans—nowhere close to the RAND-calculated requirement of 20 per 1,000 citizens as having been pointed to by the Balkan experience.

Problems in Afghanistan include an initial planning/strategy breakdown, with no standing mechanism in place to ensure adequate planning and interagency coordination; a months-long fight over appointing an Afghanistan coordinator; inability of aid and reconstruction workers, election officials, and commercial interests to enter large parts of the country because of poor security; rebounding poppy production; resurgence of Taliban and al-Qaeda; warlords' continued control of most of the country outside of Kabul; and the undermining of Karzai's government, authority, and legitimacy by our cozy relations with warlords and use of proxies. In sum, we tout the existence of government and the passing of the constitution as successes, but we essentially have only the trappings of government and the rule of law without building effective state institutions that will uphold them.

So far as the situation in Iraq is concerned, the planning/strategy failures are legion: there was no plan to prevent looting (which was ignored), no plan for security/stabilization, no plan for running the country; bureaucratic warfare between the Department of State and DoD over team personnel and other issues was not reined in by the National Security Council; Presidential Envoy Bremer's Coalition Provisional Authority and its predecessor "have been undermanned and operating with Team B from the beginning," with no standby capacity and bodies having to be scrounged from the State Department; the international police, justice, and rule of law teams were never brought in; there was no planning for a stabilization force, there were not enough troops, and there was no mandate to perform stabilization tasks for what troops we had; the war has essentially continued throughout the reconstruction effort, with lack of security shutting down some contractor work; there is a lack of good intelligence; it is not clear to whom the government will be transferred, regardless of whether the June 30, 2004, deadline is retained; public information/psyops needs fixing; lack of greater international involvement hurts the legitimacy of our effort.

The military has performed admirably under strained circumstances, and it is very good at institutionalizing lessons learned, "but asking the military to take on the enormous reconstruction burden without sufficient manpower, training, or civilian counterparts is not a long-term strategy for success."

Mr. Jaque Grinberg.

The UN, according to Mr. Grinberg, "has a crucial role to play in peacekeeping and nation-building." Its reputation in some quarters is considerably worse than its objective record. In his analysis, three themes are particularly relevant.

First, peacekeeping is here to stay, and we should therefore review our diverse assets objectively. The UN has far more to contribute in these endeavors than might be apparent. It has unique legal authority and is the prime instrument for conferring legitimacy. It also has unique moral authority, with the UN Charter still the primary document encapsulating the collective aspirations of mankind. No party to a dispute has ever disputed the message of the Charter. Only the UN has a global membership from which appropriate assets can be gathered to form precisely tailored responses to the infinitely multifarious challenges that arise. Finally, the UN offers the advantage of visible group burden sharing, thus promoting a sense of equity and solidarity.

Of course, no one can say that the UN is a flawless vehicle. In the last 15 years, all parties have had to climb a steep learning curve in handling international disorder. The UN had a head start and does some things better than other civilian organizations. In judging the UN's assets and performance, one should take care to distinguish between its role as an operational organization that performs missions around the globe, and its role through the Security Council as a legislative and policy body.

The second theme is the need to adjust policy and operational settings in approaching peacekeeping and nation-building endeavors. Several persistent problems have continued to elude satisfactory resolution (and these generally apply regardless of sponsor—UN, coalition, or national). There is a pressing need to improve civilian planning for peacekeeping operations, so that "every time doesn't seem like the first time." We must also resolve the conundrum of dealing with pressure to define an exit strategy, while at the same time maintaining and expanding local commitment and cooperation with the peacekeeping force. It seems better to identify realistic and understandable goals, but not put a definitive time limit on the intervention. Closely linked with the foregoing concern is the challenge facing democratic societies of maintaining the domestic commitment beyond their 3-to-4-year election cycles.

Still another problem is how to handle the numerous peacekeeping "gaps": the local leadership gap; the security gap; the rule-of-law gap; the jobs/expectations gap. Finally, there is the problem of reconciling the efforts of international financial institutions, development agencies, and national political priorities. The problem grows out of multiple diverging timetables and funding sources, both between organizations and within governments.

The third and final theme consists of a snapshot of the UN's vast and expensive peacekeeping mission in the Democratic Republic of the Congo (DRC). With an area of well over two million square kilometers, the DRC is a huge country about a quarter the size of the United States. It is also an economic powerhouse, with rich mineral resources (gold, diamonds, copper, and 70 percent of the world's cobalt deposits), 50 percent of Africa's hardwoods, and 10-12 percent of the world's hydroelectric capacity. But it is also a vast cauldron of human misery, with a long legacy of war, 3.5 million dead, 17 million with marginal food, 1.3 million infected with HIV, 1,010 dead from land mines, 3.4 million internally displaced persons, 350,000 refugees throughout Africa, and a ranking of 167th out of 175 nations on the UN Human Development Index.

The UN peacekeeping mission in the DRC is among the largest and most complex the world body has ever undertaken, with a $667.27 annual budget, 2,792 civilian personnel, and 10,700 military personnel. The number of UN member states contributing people is 109, with 52 of those contributing military personnel. The troop contingent strength is 9,940, plus 558 military observers and 202 staff and liaison officers.

There are grounds for optimism, revolving around a growing national identity, war fatigue, a potential peace dividend, the interests of Congolese leaders, a changing regional dynamic, and international commitment. Mr. Grinberg concludes, nonetheless, that "the jury is still out." The UN presence has been a determining factor in saving thousands of lives and giving hope for an end to "the world's greatest living tragedy." If the UN succeeds in the DRC, there is a fair chance it will have created the basis for stabilization and development in the entire Central Africa sub-region.

PANEL V: *DEMOCRATIZATION*

Professor Daniel Brumberg.

In his November 6, 2003, speech before the National Endowment for Democracy, President Bush laid out an ambitious vision for a "forward strategy of freedom in the Middle East." Though some observers attributed the speech to the White House's desire to redefine the purpose of the Iraq war, in fact the administration's neo-conservatives long have argued that the toppling of Saddam Hussein was the first shot in a long campaign to democratize the Middle East. Bush's speech showed clearly that he has embraced the conviction that it is the "calling" of the United States to extend the global democratic revolution to that area of the world.

Of course, it has been the practice of U.S. presidential administrations over the last 10 years to pursue a liberalization strategy in the region whereby various democracy aid programs were employed to encourage the emergence of political rights and civil liberties through civil society organizations. According to Professor Brumberg, of the 18 Muslim-majority Middle East states, "ten can be classified as liberalized, or partial, autocracies" (Algeria, Bahrain, Egypt, Jordan, Kuwait, Lebanon, Morocco, Oman, Qatar, and Yemen). Unlike full autocracies, whose survival depends on quashing all political competition or opposition, the longevity of liberalized autocratic regimes depends on their acquiescing to, or even promoting, a measure of managed political openness and pluralism. Thus the American strategy of reinforcing civil society organizations in the hope that eventually they will push ruling elites to move beyond state-managed political liberalization is fully consonant with the regime survival strategies of Arab rulers. As the late Egyptian president Anwar al-Sadat once put it, "Democracy is a safety valve so I know what my enemies are doing."

It should be noted, of course, that "no serious democratization strategy has the slightest chance of success so long as the Palestinian-Israeli conflict continues to simmer and periodically boil over." Until the administration shows in deeds that the Palestinians are as deserving of independence and democracy as the Iraqis, "all talk of democratization in the Middle East will ring hollow in the minds of Arab youth."

Setting aside the case of Iraq for a moment, for the United States to go beyond its long-time strategy of encouraging piecemeal liberalization as a means of eventually achieving something like genuine democracy and representative government, it would have to tackle directly the key obstacle to democracy: the institutions and ruling ideologies of Arab states. Such a shift will be very difficult to achieve since it will require pressuring some of the same Arab leaders whose support the administration needs in the war on terrorism. Moreover, the administration would have to tread delicately and wisely in its choices of which countries to attempt to democratize, because it could conceivably open the door to Islamic hardliners capable of capitalizing on the instruments of democratic governance to create a radical Islamic state.

Silver Spring , MD 20910-2254
U.S.A.

To:
ABE APTP ORDERS
800 AVONDALE AVE
QMN9AM74
COLUMBUS OH 43212
U.S.A.

FOR CUSTOMS ONLY

Shipping Manifest

Phone: 0000000000

Sales Order No.: 774829929
AbeBooks Purchase Order No.: 738420679
Processed: September 16, 2026 **Delivery Time:** 5 - 20 business days

Item	Author	Title	Bookseller Inventory No.
1	Matthews, Lloyd J.	Winning the War by Winning the Peace: Strategy for	91635

Book Description: vii, [1], 53, [3] pages. Conference Report from the 15th Annual Strategy Conference held in Carlisle Barracks, PA April 13-15, 2004. Some wear to the cover edges. The Strategic Studies Institute (SSI) is part of the U.S. Army War College and is the strategic level study agent for issues related to national security and military strategy with emphasis on geostrategic analysis. Studies produced by civilian and military analysts concern topics having strategic implications for the Army, the Department of Defense, and the larger national security community. Paraphrased from the Foreword by Douglas C. Lovelace, Jr., SSI's Director: During each of the last 15 years, the U.S. Army War College has sponsored a strategy conference that addresses a major security issue. The conference theme for year 2004 was "Winning the War by Winning the Peace: Strategy for Conflict and Post-Conflict in the 21st Century." Informed by the Korean, Vietnam, and Gulf wars fought by the United States and its allies during the last half of the 20th century wars in which, despite the qualitative superiority of our forces, the outcomes proved to be less than satisfactory the conference theme for 2004 entailed a deep probe into the question of how can the West, in this new century of omnipresent terrorism, capitalize on its superior military and economic might to achieve a satisfying and enduring modus vivendi. conclusions. The author is a retired U.S. Army colonel. His military assignments included command at platoon, company, and battalion level; advisory duty in Vietnam; editorship of Parameters, the Army War College quarterly; and the associate deanship of the U.S. Military Academy.

Please retain this packing list for your records. If you have any questions about this order, please contact the bookseller using the Ask a Bookseller a Question link available in your AbeBooks account.
**Please Note: AbeBooks has collected payment for this order.

Thank you for shopping with AbeBooks.com.

AbeBooks

The most likely candidates for a successful democratization strategy thus would be those who already boast an electoral system of such independence and competitiveness that, if and when Islamists do enter a genuinely open election, they must face, negotiate, and ultimately share power with non-Islamic parties. Morocco is one of the few countries in the Arab world where the risks of full-blown democratization might be worth taking. The leaders of Egypt, Jordan, Kuwait, and Bahrain would be very hesitant to embrace any strategy that moves them much beyond the limits of present liberalization.

The Bush administration has chosen to forgo the path suggested for Morocco or a similar Arab state. Rather, it has hitched all its hopes on a policy of military intervention and forcible regime change in Iraq. Its guiding assumption appears to be that the legacy of authoritarianism is so deeply embedded in the social, economic, and political soil of the Arab world that there can be no hope of genuine political reform unless one prominent Arab country is given the chance to demonstrate for the entire region how to "get it right from the very beginning." One can readily imagine that such a fortunate outcome might eventually inspire Arab leaders elsewhere to negotiate democratic solutions with their opponents.

However, the experiment in Iraq may eventually produce the opposite result from that intended by the United States—more rather less autocracy in the Arab world. In Professor Brumberg's view, the chances of avoiding such a result are diminishing because to prevail over the long term, any credible power-sharing arrangement between the Kurds, Shiites, and Sunnis requires, among other things, "a long-term political and especially military commitment from the international community." Unless the United States gets lucky and somehow achieves such a commitment, there is a high risk that the Shiites and Sunnis will drift into a temporary marriage of convenience under the nationalist banner (as they did in 1920 against the British) to expel the American invader. Viewing the resulting chaos in Iraq, many Arab leaders, both incumbents and those in opposition, would then decide to "let the present flawed system totter along since trying to fix it would likely only make the situation worse."

Dr. Larry Diamond.

Anyone who has been watching or reading the news must know that things have been slipping badly in Iraq. Usually, the mass news media exaggerate the negatives and suppress the positives. We have not heard nearly enough about the good work that our mission there—the Coalition Provisional Authority (CPA)—has been doing to rebuild the country and its infrastructure and support democratic institutions, organizations, and values. But, unfortunately, the bad news indeed has been quite bad. Fallujah is not of overriding concern. The news there has been shocking and horrific, but this is a limited uprising from a minority section of the country. It does not threaten the overall viability of the political transition program in Iraq.

The Shiite uprising in the south is another story, however. Many scholars and historians of Iraq long have warned that an uprising among the Shiite population would spell doom for the Coalition and for any hope of a peaceful transition to anything resembling democracy. No doubt this is true. But we are not yet facing a generalized Shiite resistance. Rather, we are locked in a confrontation with a ruthless young thug, leading a fascist political movement that is using religion in a twisted way to achieve its own crude ambitions for power.

For the past year, we have been engaged in a difficult effort against great odds, to rebuild Iraq and foster a transition to democracy in that unfortunate country. It has been a turbulent road, littered with mistakes on our own part and obstacles posed by Iraq's authoritarian neighbors, who are panicked at the thought of having an emerging Arab democracy (or, in the case of Iran, a predominantly Shiite democracy) on their own border. But we nevertheless have seen some inspiring progress.

A variety of civil society organizations and think tanks are emerging and finding their legs. The Iraqi Governing Council has adopted an interim constitution which will distribute sovereign power in Iraq during the period from the transfer of authority to an Iraqi government on June 30, 2004, to the seating of a new elected government under a permanent constitution by the end of 2005. Around the country, local and provincial councils have been formed with varying degrees of popular involvement and support. In many cases these have achieved broad participation and consultation (and in a few, even direct elections) that have produced local governments with far more legitimacy than anything that preceded them. A vigorous campaign is underway to educate Iraqis about basic democratic principles and to draw them into dialogue.

There are serious problems with the interim constitution. Iraqis complain that there was too little popular involvement in the process, and the document gives far too many veto rights to the Kurds and other minorities. They worry—quite mistakenly—that it will allow the United States to conclude a binding treaty with an unelected government. Some wish that the constitution was rooted more exclusively in Islam. But still, with all of the problems and controversies, the transitional law is a step forward for Iraq, and gives that country "the most progressive and liberal basic governance document of any country in the Arab world."

Preparations are underway to bring the political promise and timetable of this constitution to fruition. As a result of negotiations conducted by UN representative Lakhdar Brahimi in February 2004, a compromise agreement was reached providing for direct election of a transitional parliament by the end of 2004 or January of next year. Ambassador Brahimi is negotiating a framework for the Iraqi interim government that will assume power on June 30, 2004. A separate UN team led by its chief elections expert, Carina Perelli, is in Iraq consulting on and helping to define the structure for an independent electoral commission that will administer elections in the country, and on the crucial question of how to structure the electoral system that will select members of the transitional parliament.

However, the political transition has been heading into a storm of violence and intimidation. The biggest cause for concern in this regard lies with the Shiite heartland of Iraq. Over the past year, a growing array of armed private militias—loyal to political parties and religious militants, and funded and encouraged by various power factions in Iran—have been casting a shadow over the political process. While we have been focusing on building civil society, educating for democracy, writing a basic law, and negotiating the future structure and timetable of transitional government, the militias have been building up their weaponry and recruiting fighters. The total number of these militia fighters in many provinces well exceeds the combined strength of the new Iraqi armed forces.

Their armies have been stocking heavy arms, intimidating opponents, and preparing for the coming war in Iraq. This war will not fundamentally be a war against American occupation or international domination, though now that it has broken out prematurely, its first phase is being framed and justified in those demagogic terms. It is a war for something more primal: the acquisition of power.

Some in the CPA who have seen this threat gather in recent months have warned that, unless the pseudo-religious militias are demobilized and disarmed—through negotiation, ideally, but through force where necessary—a transition to democracy in Iraq will become impossible. Rather, at every step of the way, from parties canvassing for supporters to the registration of voters to the election campaign to the casting and counting of votes, the democratic process will be thwarted by strong-arm methods and fraud, and the quest for a free and fair political process will fail.

Fortunately, key officials within the CPA have quietly seized upon the issue of party militias as a priority. Over the last 3 months, a plan has been prepared and negotiated for the comprehensive disarmament, demobilization, and reintegration of all the major militias. Using the lessons of other negotiated programs, this plan will offer members of the various militias generous financial incentives to lay down their private arms, disband, and be integrated either into one of the new Iraqi armed forces, such as the army, civil defense corps, or police, or into other sectors of the civilian economy, with training if necessary to prepare them accordingly.

Experts have been developing and negotiating this disarmament plan with key national and local militia leaders. They believe they are making significant progress. But in this kind of effort, it is very nearly all or nothing. No party militia, Dr. Diamond emphasizes, "will disband if it believes it will risk collective suicide in doing so." Party militias probably have an offensive intent: to ensure that they can seize by force and fraud what they cannot be confident of winning fairly at the ballot box. But they also have a defensive intent: to prevent such fraud by their enemies, to protect the lives of their leaders, candidates, and campaigners, and to maintain order in areas they control. Unless its rivals are demobilized at the same time it is, no serious militia will sincerely cooperate.

The most dangerous impediment is Muqtada Sadr, abetted by a ruthless militia, which stands outside any process of negotiation and voluntary disarmament. A fiery 30-year-old mullah whose father and brothers were martyred in the Shiite resistance to Saddam, Muqtada has nothing of the Islamic learning and sophistication that would put him anywhere close to the religious stature and authority of an ayatollah. But he knows how to organize, mobilize, and intimidate. He has used the reputation of his father among the poor urban masses, and the language of historical resistance to external impositions to mobilize a growing following among young downtrodden urban men, in particular. His support is confined to a small minority among the Shiites of Iraq, but it is the kind of minority, demographically, that "makes revolutions and seizes power," and its devotion to his declarations and obedience to his commands are strong.

As with the Nazis, Muqtada has been guilty of brazen crimes well before his effort to seize power openly. A year ago, Sadr's organization stabbed to death a leading moderate Shiite cleric, Ayatollah Abdel-Majdid al-Khoei, who would have been a force for peaceful democratic change and a dangerous rival to Sadr. The murder took place in the Imam Ali mosque, Shiite Islam's holiest shrine.

Recently, Sadr's propaganda, both in his oral statements and through his weekly newspaper, the *Hawza*, has become increasingly incendiary, propagating the most outrageous lies (for example, that the United States was responsible for recent deadly bombings) deliberately designed to provoke popular violence. Finally, on March 28, 2004, after months of costly delay, the Coalition began to move against him. Ambassador L. Paul Bremer ordered the closure of the newspaper and ordered the arrest of those responsible for the murder of al-Khoei.

Sadr responded to these arrests by unleashing what can only be described as the beginning of a revolutionary campaign to seize power. Having already occupied numerous public building in recent months, his followers took over the offices of the Governor of Basra and assaulted police stations in several cities, including Karbala and Najaf, with their sacred Shiite religious shrines.

In response, Ambassador Bremer declared Muqtada Sadr an "outlaw." Now there is no turning back. If any kind of decent, democratic, and peaceful political order is to be possible in Iraq, the Coalition will need to crush Muqtada's attempt to seize power by force, and dismantle his Mahdi army.

This will only be the beginning of a campaign to control privatized violence and construct a rule of law in Iraq. Many additional forceful measures will be needed. These must include vastly enhanced security on Iraq's borders with Iran, Syria, and Saudi Arabia, and blunt messages to the Iranian and Syrian regimes. The Iranian mullahs must understand that we will not stand by and watch them brazenly subvert the quest for democracy in Iraq. The Iranians must be made to know that two can play this game. They face a mounting resistance to oppression in their own country, and we can do much more to assist that resistance, in direct proportion to the irresponsibility of their actions in Iraq and to their disregard of international strictures against nuclear proliferation. The same goes for Syria, through which—Iraqi democrats are convinced—al-Qaeda terrorists are passing freely into Iraq, with the assistance of Syrian intelligence.

At best, we are in for a rough period in Iraq. Tragically, there are going to be many more Iraqi, American, and other Coalition casualties. As the violence increases, more and more critics will be uttering the words "Vietnam" and "quagmire." But this is not Vietnam—most Iraqis are disgusted with the violence and clearly want the chance to freely elect their own government.

There is only one way out of the dilemma we confront, and it is not via retreat or "holding the line" at an untenable status quo. Instead, it is to move forward assertively, to commit all the troops and resources it will take to defeat the religious bolsheviks and common thugs, to build up the security elements of the new Iraqi state, and to give Iraqis a chance to democratize free of intimidation. Such renewed military resolve must be combined with a political strategy to produce a significantly more inclusive and representative interim government than the 25-member Governing Council with whom we have worked since the middle of last year. Only if we have a political strategy to draw in Ayatollah Sistani and the wavering bulk of the Shiite population can we prevail in this campaign.

According to Dr. Diamond, President Bush is right to insist that the rising tide of violence will not shake our resolve, or alter our plan to transfer authority to an Iraqi interim government on June 30. Any delay in that transfer would further inflame the situation and feed suspicions that we are bent on permanent dominion in Iraq. But the Bush administration has not leveled with itself or the American people about the resources that will be needed to achieve any kind of victory in Iraq. Our mission in Iraq has been under-resourced from the start. We do not have enough troops there. We do not have enough secure transport, including helicopters. And we do not have enough armored cars, trained security personnel, and high-quality body armor to protect the many civilians—career diplomats, aid workers, and term consultants—who have been risking their lives for this cause.

In the next few months, we could lose the new war for Iraq if we do not project the necessary resolve, combined with the right political strategy to generate a more inclusive and legitimate government. If we can neutralize the militias while building up the new Iraqi instruments of law, a small miracle could yet unfold by January 2005: reasonably free and fair elections for a transitional government (which will also write a new permanent constitution). Then, what now appears to be a downward spiral into civil war could well be averted, and "this long-suffering country could be placed on a rocky but realizable path to democracy."

Professor Andrew Reynolds.

Stating flatly that there is not a military solution in Iraq, Afghanistan, or Sudan, Professor Reynolds goes on to aver that only political measures abetted by military containment and neutralization of armed opposition will work. The primary instrument in achieving such a political solution is a national constitution. The art of constitutional design can be elaborated by resort to a series of medical analogies.

First, "do no harm." Constitutional engineering has lain at the heart of political settlements as diverse as the Dayton accords for Bosnia, the transition from apartheid in South Africa, the power-sharing design for Fiji, and Northern Ireland's second try at uniting the two sides in viable association. In those parts of the world where ethnic plurality prevails, the design of political institutions, including constitutions, has been seen as a useful tool "to encourage socio-political stability and to promote inter-ethnic accommodation." Institutions must be inclusive, bringing under their ample folds all groups, leaving no outsiders. Getting the rules right, just as with getting the right medicine, is crucial. Particular attention must be paid to democratic design, that is, shaping the mechanisms that give effect to democracy and representative government.

Just as doctors seek to diagnose and treat a sick or bleeding patient with a variety of drugs and behavioral modifications, the constitutional physician looks on an ailing society and attempts to determine what institutional medicines might best stem the blood flow and provide the long-term foundation for a return to health and vitality. Poor diagnosis leads to inappropriate treatment, but one can understand the prevalence of diagnostic error in view of the symptoms of inter-ethnic violence and hostility as well as state failure that could also result from other pathologies.

Constitutional design should reflect structural analogies to the medical continuum of triage, emergency medicine, convalescence, and longer-term health management. What may be good for immediate conflict crisis management (or triage) may not be good for longer-term care and consolidation of a democratic political order. For example, the interim constitution for Iraq, written in the context of war and insurgency, will not be the most appropriate instrument for the eventual independent, democratic Iraq. Any healthy constitution must contain mechanisms for changing it in the light of future need.

Like medicine, "governance must be approached holistically." It is a mistake to treat a specific ailment (such as disputed elections) while failing to engage in a broader diagnosis which seeks to find and treat those underlying causes that may lie elsewhere in the body (e.g., addressing the accountability of those in power, the efficacy of the judiciary, etc.). Moreover, institutional drug prescriptions can work against each other, thus exacerbating the illness, if the elements of the treatment regimen are not complementary.

Professor Reynolds warns against a "rush to surgery—or to elections." They can be traumatic, thus only a relatively stable patient is a good candidate. In the absence of reasonable national stability, elections represent a threat to any peace process. For example, in Afghanistan, only 2.3 million of 10 million voters have been registered. Most are male Pashtuns concentrated in the same area. These deficiencies result from the "logistics of security." A satisfactory election can hardly flow from such a skewed electorate. In many post-conflict situations, the infrastructure to facilitate legitimate elections is lacking, and a climate allowing for unfettered campaigning often is nonexistent. Moreover, doctors should not discharge the patient prematurely. All too often the first election is seen as the end goal, with the doctor then discharging the patient before any of the structural or civil foundations of a multiparty electoral democracy have had time to take root (think of Liberia, Cambodia, and now Afghanistan).

Finally, it must be remembered that even the best medicine has its limits. Just as medicine cannot save the terminally ill, an excellently crafted set of political institutions cannot save a truly sick society. The soundest constitution can be overwhelmed by disintegrating security. Political institutions are just one part of the puzzle. Other instruments of the state need to be in good working order. For example, a fully functioning judiciary, a progressive education system, high levels of employment, energetic economic development, and internal security are all essential to provide the foundation of a healthy polity.

Dr. Lawrence Korb, Luncheon Speaker.

In conducting the global war on terrorism, the United States has essentially three strategic choices, in the opinion of Dr. Korb. The first is preventive war, a course usually associated with the neo-conservatives and hardliners. It asserts the right to attack unilaterally known enemies at times and places of our choosing in order to forestall future attacks by depriving the enemy of the wherewithal to conduct them. It is not to be confused with anticipatory self-defense, which suggests specific and certain intelligence of an imminent enemy attack in the process of initiation. Rather, it is a generalized response, based on the enemy's credibly declared intent to attack and continue attacking, and on the enemy's presumed possession of WMD which, if used, would inflict massive, grievous, and irremediable harm. Rather than simply waiting and attempting to intercept specific attacks once set in train, or attempting to deter such attacks by threatening to retaliate after the fact, the preventive strategy goes further by asserting the right to minimize the possibility of any future attack by destroying the enemy at his source, even if it means attacking the states that harbor him.

Dr. Korb finds two corollaries emerging from the preventive strategy: the United States will maintain military dominance in the world regardless of cost (possibly a message to China as well), and it will work to make the world democratic (an important reason being to get rid of terrorists).

The second strategy that America might choose is *realpolitik*, or political realism. *Realpolitik* is no less given to an appreciation of power than the preventive strategy, but "it believes that deterrence and containment still work," even against rogue states and those that sponsor terrorism. Though no enemy of democracy, the school of realpolitik, where forced to choose between democracy and stability, would opt for the latter. To attempt to impose democracy around the globe would overextend our military and economic power as well as our political patience. There is a real risk that the American people, faced with the prospect of indefinite struggles around the globe to impose our vision of political salvation on other peoples, would resign from the game, insist on bringing our forces home, and withdraw into a cocoon of isolation and passivity.

The third potential strategy goes under several names—cooperative security, liberal internationalism, multilateralism, etc. Certainly it recognizes the global threat posed by the presence of WMD in terrorists' hands. But its primary response would be to work diligently to improve the economic and social conditions in states that breed the terrorist mentality. Meanwhile, if we are forced to take military action, we should do so through the UN or under the umbrella of other suitable multilateral institutions. A fundamental and enduring solution, however, can be found only in economic and social redress, following which political alienation would subside.

A fourth strategy, of course, would be to mix and match from among the three, fashioning a composite strategy that borrows the best from each and tries to avoid the worst.

With regard to the situation in Iraq, which flows from our present proactive strategy, we must have more boots on the ground. We should therefore increase the active Army by at least two divisions and move more Military Police and Civil Affairs units from the reserves into the active force. Even if we formed a corps of reconstruction specialists apart from the Army, we would still need an increase in the number of troops. When the U.S. push to launch an attack against Iraq was being debated before the Security Council, we probably could have salvaged a greater degree of multilateral involvement, and thus legitimacy, had we worked at it longer, but it is impossible to know this for sure.

PANEL VI: *IMPLICATIONS FOR MILITARY CAPABILITIES AND FORCE STRUCTURE*

Dr. Daniel Goure.

Winning wars and winning the peace are not the same thing. They require different plans, capabilities, and skills, and they operate on different timelines. The argument that "the military is the institution best suited to performing peace and stability missions is not self-evidently true." Granted, it has people, equipment, organization, and some training, but it is not clear that much of what we have is applicable to the problem. The military's capabilities, including those of large troop formations, may be relevant, but they are also extremely expensive, scarce, and dwindling commodities.

U.S. forces have been reduced by 40 percent since 1990, while the demand on those forces during the same period has more than doubled. It is one thing to propound peace operations when we have 18 Army divisions (which we had in 1990), but quite another to propound them when we have only 10 as today. Allied reductions have been even more severe, with their military budgets having shrunk more than that of the United States. Their forces essentially are unusable, barely able to handle even small-scale contingencies. Given the number, scale, and duration of commitments, plus the lack of substantial help from allies, American forces are over-stressed to an unsustainable degree.

Thus the U.S. military can either "fight the nation's wars, or achieve the peace." It cannot do both. We need to stop looking to the military as the means to perform missions that are primarily political, legal, economic, social, and psychological in nature. Successful peacekeeping and nation-building efforts are the exception, not the rule. The military can support such efforts, but it cannot lead them and should not provide the bulk of the resources. The considerable time it takes for successful nation-building cannot be reconciled with the design and deployment tempo of the present force.

For the global war on terrorism, the United States needs a different kind of force, though the precise nature of the changes needed is not clear. Modular light infantry forces integrated with intelligence, communications, civil affairs, and psychological operations seem indicated. Mobility, force protection, tactical intelligence, and superior firepower are the critical requirements. Mobile light forces are needed to hunt down insurgents in difficult terrain. Infantry is needed for close combat in urban terrain. The Israelis, however, have shown that heavy combat forces also do well, and actually may be safer than light forces.

Our organization for counterterrorism is inadequate, with the CIA too light and the military too heavy and regimented. A new arm of the CIA is worth considering, one that can integrate human intelligence, clandestine operations, and field operations. Armed contractors would be useful in certain missions, while our allies, even with their depleted forces, could be highly useful in certain niche and specialty roles.

Colonel Richard Hooker.

With the enormous human and material investment the United States is making in military transformation, it is vital that we get it right and not fall victim to our own hype. There is much to admire about the goals of force transformation: it plays to American strengths in technology and engineering; it increases our capacity to apply lethal force while reducing the prospect for suffering casualties; it aims toward a reduced requirement for ground troops; it promises short, sharp campaigns; it does not necessarily rely on allies, who might prove uncooperative or unreliable. There is an element of risk, however, in embracing the transformation agenda entirely without noting and addressing its potential shortcomings.

Transformation is focused on technology and the notion of the networked information grid. It envisions an interconnected sensor grid able to pass information and intelligence instantaneously to firing platforms. In theory, this grid will provide full situational awareness to commanders, who can then select and attack the most critical and vulnerable target sets for maximum effect. Information superiority, enabled by systems that can relay data seamlessly from sensors to shooters, thus produces faster decision cycles; forestalls enemy reactions; creates more friendly options; and minimizes risks. After several years of transformational work, however, the effort to put these concrete capabilities into the hands of actual warfighters deployed on the ground "has not progressed much beyond the powerpoint stage."

A basic flaw in transformation thinking is a misconception about the nature of war itself. Transformation proponents insist that certainty can be approached in war. They assume that war is a coldly logical process of action and reaction, and thus that given sufficient data, the enemy's intentions can be calculated. They lose sight of the facts that war is a clash of wills, the relative strength of which will always lie hidden until after the fact, and that commanders are not purely rational automatons but rather human beings, blessed or cursed with all the infinitely variable passions to which man is heir.

Rejecting the factors that have always rendered war so notoriously uncertain, the proponents of transformation anticipate not only the ability to see the enemy at all times and in all places, but to actually predict opposing moves, even when the enemy commander may not himself have decided yet what he will do. In the view of transformationists, full situational awareness will thus largely, if not completely, dissipate the fog and friction of war.

The foregoing is a dangerous and unwarranted position for soldiers to take. The essence of this vision of future war is to reduce action on a battlefield to the level of a targeting drill. This technicist view reflects the experiences and intellectual predispositions of many transformation advocates with air and naval backgrounds. Their arguments reveal few conceptual distinctions between the levels of war; little understanding of war's intensely human character; insufficient recognition of differences between the various warfighting domains of land, sea, air, and space; and a misunderstanding of service core competencies and their contributions to joint warfare.

Uncritical acceptance of the notion that transformation constitutes a broad panacea for the problem of waging war leads to muddled thinking on almost every level. Official publications, for example, purport to see a nexus between nonstate actors like al-Qaeda and an urgent need to embrace network-centric warfare. But net-based war was first articulated years before 9/11 and is clearly more suited to attacking fixed modes and targetable centers of gravity than small cells of widely spread and loosely organized terrorists who communicate by messenger and encrypted e-mail.

Colonel Hooker concluded with the observation that overemphasis on airpower, precision engagement, and information superiority at the expense of an ability to seize, hold, and control ground will pose grave risks for the United States in the future if allowed to crowd out, rather than complement, other critical capabilities.

Lieutenant Colonel Robert Cassidy.

Historically, the U.S. Army has enjoyed many successes in counterguerrilla war, even including some in Vietnam. However, overall failure in Vietnam caused the Army to turn its back on that war as soon as possible and refocus on the prospect of a big war in Europe, the scenario long preferred by the American military culture. The result has been that the hard lessons learned in Vietnam about fighting guerrillas were not preserved in the Army's institutional memory. Under the mantra "No More Vietnams," the Army simply lost interest.

But since the Army and its Coalition partners now find themselves prosecuting counterguerrilla wars in Afghanistan and Iraq, it is useful to revisit the experiences of Vietnam and other counterinsurgencies involving American forces to rediscover those lessons that might be applicable in the Middle East. Lieutenant Colonel Cassidy finds that three operational programs employed by Military Assistance

Command Vietnam (MACV) show particular promise as a source of ideas for adaptation and use today:

- Civilian Irregular Defense Groups (CIDG). These groups, trained and organized by the Special Forces, were recruited from various ethnic tribes that inhabited the mountain and border areas. They conducted long-range reconnaissance and interdiction patrols along Viet Cong and North Vietnamese infiltration and supply routes; raided enemy base areas; employed hit-and-run guerrilla tactics against regular enemy units; and provided security in isolated areas.
- Marine Combined Action Platoons (CAPs). In the Marine enclave in the north, CAPs were formed consisting of a Marine rifle squad combined with a platoon of indigenous forces. Living in their own local village or hamlet, and operating from it, the CAPs destroyed local Viet Cong infrastructure; provided security and helped maintain law and order; organized indigenous intelligence nets; conducted civic action projects; and disseminated propaganda against the Viet Cong.
- Civil Operations and Revolutionary (later Rural) Development and Support (CORDS). This organization was established under MACV to unify and provide single oversight of the pacification effort. It provided support, advisers, and funding to the police and regional and popular forces, aiming to reinforce self-defense and self-government at the local level. Particular efforts, including the Phoenix program, were devoted to rooting out the Viet Cong infrastructure.

There is also a treasure trove of lessons to be drawn from the Banana Wars, the Philippine Insurrection, and the Indian Wars. The Marines' *Small Wars Manual* (1940) codifies their experience in a particularly useful manner, and there are high-quality works by civilian scholars from which enduring principles of American-fought counterinsurgency wars can be drawn.

Lieutenant Colonel Cassidy concluded with several recommendations: allot counterinsurgency warfare a more substantial share of the U.S. Army Command and General Staff College curriculum;

include the history of counterinsurgency warfare in Army officer education from precommissioning through the senior service college attendance; publish a counterinsurgency equivalent to U.S. Army Field Manual No. 3-90, *Tactics* (2001); consider fusion or greater collaboration between the Special Warfare Center at Fort Bragg, North Carolina, the Peacekeeping and Stability Operations Institute at Carlisle Barracks, Pennsylvania, and the Combat Arms Doctrine Development and Force Modernization Office at Fort Leavenworth, Kansas; and consider the creation of a modular civilian civil and rural development corps subordinated to the commander of the joint task force during counterinsurgency campaigns.

Brigadier General Ryan, Closing Address.

Brigadier General Ryan briefed the Army Campaign Plan for "transforming" the service, which itself is part of the Army Plan. He first cites the mission: "Build a campaign-capable, joint, and expeditionary Army in this decade to provide relevant and ready landpower to combatant commanders and the joint force while sustaining operational support to combatant commanders and maintaining the quality of the all-volunteer force."

A summary notion of the comprehensiveness of the plan and the topics it addresses was conveyed through the following descriptive rubrics: Security Environment; Nesting Army Strategy; Strategic Planning Guidance Overview; Relationship (among National Security Strategy, Defense Strategy, National Military Strategy, and Secretary of Defense's Strategic Planning Guidance); Changing Operational Environment; Major (Anticipated) Changes; Reducing Deployment Tempo by Building Capabilities; Army Campaign Plan—Conceptual Framework; Army Campaign Plan Assumptions; Intent; Building a Joint and Expeditionary Army with a Campaign Capability; Major Decisions Affecting Program Objective Memorandum 06-11; Brigade Combat Team (Unit of Action) and Unit of Execution Synchronization Model; Army Macro Synchronization Matrix; Brigade Combat Team (Unit of Action)-Centric Organization; Echelons and Capabilities; and Army Campaign Plan Development Timeline.

Brigadier General Ryan explained that the term "expeditionary" in current Army parlance carries its traditional meaning but implies "a rotational pattern of deployments." Implicit in the Campaign Plan is the requirement that Army forces be able to "support two major combat operations simultaneously, one of which can be escalated to decisive victory." The fighting force itself must become more cohesive, agile, and combat-ready, but within the context of a more stable, predictable lifestyle for Army members and their families. A reduced deployment tempo will be achieved by producing additional deployable active Army brigades (brigade combat teams/units of action) through such measures as restructuring present brigades and converting MOSs. Additionally, more high-demand units such as Military Police and Civil Affairs units will be formed. But such changes will be made "while the engine's running." It will not be business as usual. The aim is to reduce the percentage of active component brigade combat teams (units of action) deployed overseas from about 67 percent of the brigade pool at the end of 2004 to roughly 33 percent by the end of 2007. Brigadier General Ryan cautioned, however, that full execution of the transformation campaign plan depends heavily upon assumed supplemental budgeting.

FINDINGS AND RECOMMENDATIONS

The conference architecture did not provide for a canvass or survey of participants' views on particular issues, so it is not feasible to arrive at consensus findings and recommendations. However, informal analysis of remarks by panelists and other speakers reveals such strong recurrent endorsements of some proposals that it will be useful to highlight them. While no claim to unanimity can be made regarding these positions, it can be said that a number of participants expressed support for them in one guise or another, and that few if any voices were raised in opposition. The positions in behalf of which significant support was expressed are as follows.

- **More troops are required for the war in Iraq**. By a wide margin, this was the most frequently and emphatically voiced finding, often expressed in the soldier's phrase, "We've got to get more boots on the ground." In fact, many presenters spoke of the troop increase not as a need, but as a foregone conclusion: "There *will* be a troop increase." And in stating this, they were not referring to Secretary Rumsfeld's emergency increase of 20,000 soldiers announced several weeks ago.

- **The date September 11, 2001, marked a historic juncture in America's collective sense of security**. Our presumption of invulnerability has been irretrievably shattered.
- **We need greater international participation in the Iraqi nation-building effort, preferably under the aegis of the UN**. Significantly, not a great deal was offered about how this step can now be achieved, although most participants appeared to favor greater multinational involvement.
- **As soon as possible, we need to turn the reins of Iraqi government over to an indigenous entity that will be viewed as legitimate and that will have the muscle to maintain order**. However, it seemed to be assumed that for an indefinite period, a Coalition troop presence would continue to be necessary in an overwatch posture, even after an Iraqi government takes formal control.
- **A finding corollary to the foregoing was that the U.S. Government, under whichever political party, must summon the will to stay the course**. To pull out prematurely, with Iraq still unstable, would be a catastrophe of unimaginable proportions.
- **Even with some troop increases, the U.S. strategy in Iraq will continue to be a moderate, hold-the-line approach until the presidential election of November 2004**. After that, decisive changes can be expected.
- **The only credible institutions for mounting nation-building efforts are the Army and the Marine Corps, but even they are maladapted for such work**. Faced with this dilemma, several conferees were inclined to explore the concept of constituting civil reconstruction corps (modeled after CORDS in the Vietnam War?) subordinate to the joint task force commander.
- **We cannot replicate Western-style democracy and representative government in Muslim-majority states of the Middle East**. This was an oft-stated theme, despite the Bush administration's declared determination to proceed with democratization. The implication was that when we propose democracy in a state like Iraq, we are, in fact, speaking of

qualified democracy, at least at the national level, with embedded structural bars to the emergence of radical Islamic rule.

- **There can be no military solution in Iraq, only a political solution**. This sentiment was expressed often but rarely explained. It seems to mean that the militarily victorious Coalition cannot successfully impose a reconstructed Iraqi government on the people by fiat and brute force. Rather, a successful reconstruction, i.e., one that is legitimate, viable, and appreciably democratic, can emerge only from the broadly inclusive participation and concurrence of a deliberative body representative of the country's political, ethnic, and religious constituencies.
- **In the global war on terrorism, conclusive victory in the classic sense is probably unattainable**. This sentiment was rarely expressed outright, but was implicit in the frequent use of such terms as "war of unlimited duration" and "war of uncertain outcome." The sentiment was also present in the view of those who regarded the best attainable result as a gradual rapproachement between the haves and have nots of the world. Here, economic integration and equality, with a consequent dissipation of alienation and mutual hostility, offered the best chance of ultimately nudging the two camps to a peaceful *modus vivendi*.

☆U.S. GOVERNMENT PRINTING OFFICE:2004-704-049/00037